M c G R
SCIENCE

Macmillan/McGraw-Hill Edition

Cross Curricular Projects

GRADE 1

Macmillan
McGraw-Hill

New York   Farmington

MW01595812

Macmillan/McGraw-Hill

A Division of The McGraw·Hill Companies

Macmillan/McGraw-Hill
Two Penn Plaza
New York, New York 10121

Printed in the United States of America
ISBN 0-02-280132-4 / 1
     5 6 7 8 9 079 06 05 04 03

Table of Contents

Plants Have Parts

Connection to Language Arts
Science helps you understand what words mean.

Match.

leaf

root

branch

seedling

All Fall Down

Connection to Math
Science helps you use math to count.

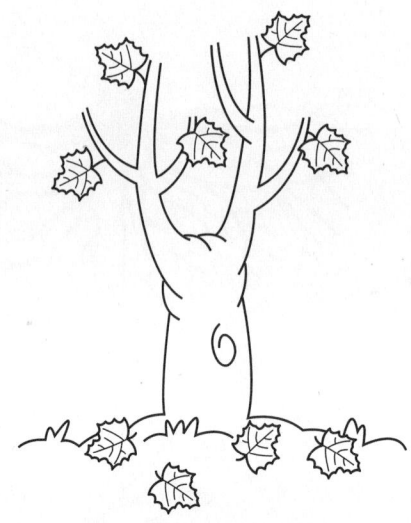

Count.

1. How many leaves are on the ground? _ _ _ _ _ _ _ _ _ _ _ _

   _____

2. How many leaves are on the tree? _ _ _ _ _ _ _ _ _ _ _ _

   _____

3. Count all the leaves. How many? _ _ _ _ _ _ _ _ _ _ _ _

   _____

4. Are there more leaves on the tree? How _____
   many more? _

   _____

A Walk in the Woods

Connection to Reading and Social Studies
Science helps you understand what you read and how to put things in order.

Read the poem. **Write the numbers 1, 2, 3 to put the pictures in order.**

I saw the squirrel climb a tree.

I saw the squirrel as high as can be.

I saw the squirrel drop a nut on me.

_____   _____   _____

- - - - -   - - - - -   - - - - -

_____   _____   _____

Name ____________________

Cross Curricular
Activity 4

What plant is it?

Connection to Language Arts
Science helps you learn new words.

Use the code.

w	a	t	e	r	m	e	l	o	n
1	2	3	4	5	6	7	8	9	10

1. The Sun helps plants grow.

   It keeps them ___ ___ ___ ___.
   1 2 5 6

2. Plants need to be wet to grow.

   They need ___ ___ ___ ___ ___.
   1 2 3 4 5

3. This plant has a trunk and branches.

   ___ ___ ___ ___.
   3 5 4 4

4. This plant grows yellow fruit.

   ___ ___ ___ ___ ___.
   8 7 6 9 10

Name _____

Plants Graph

Connection to Math
Science helps you use a graph.

Color the graph to show how many.

🌼	1	2	3	4	5

🌳	1	2	3	4	5

A Tree Changes

Connection to Social Studies
Science helps you predict what will happen next.

A tree is a plant.
Draw a tree in each season.

Fall

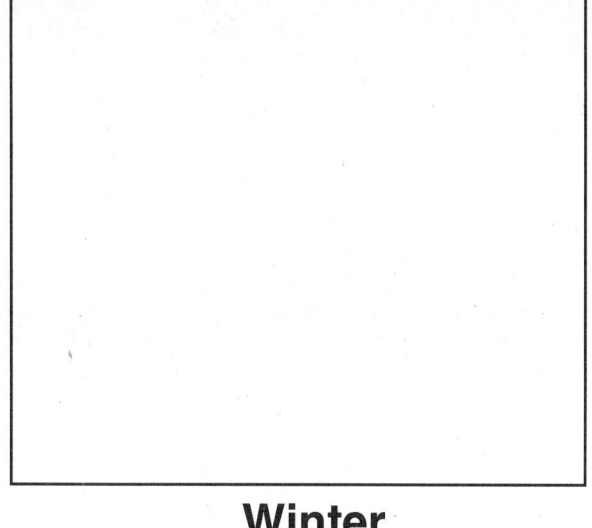

Winter

Spring

Summer

My Book of Plants

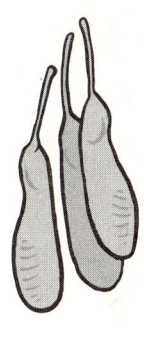
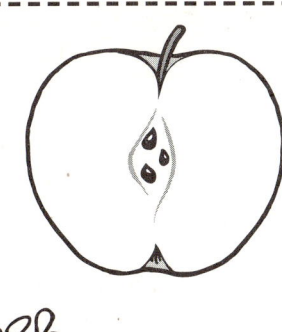

My Book of Plants

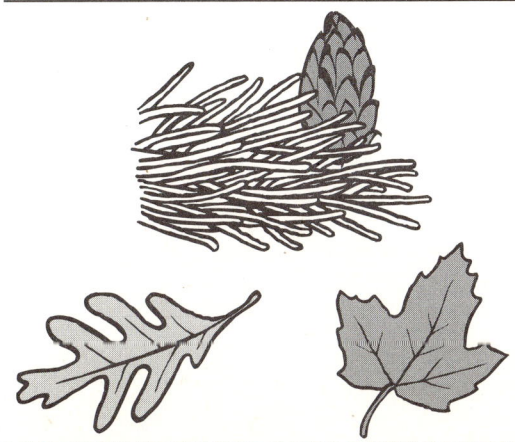

I Am a Mammal

Connection to Math and Language Arts
Science helps you put number words in order.

All mammals have hair or fur.

What is this mammal?

Follow the number words from one to twelve.

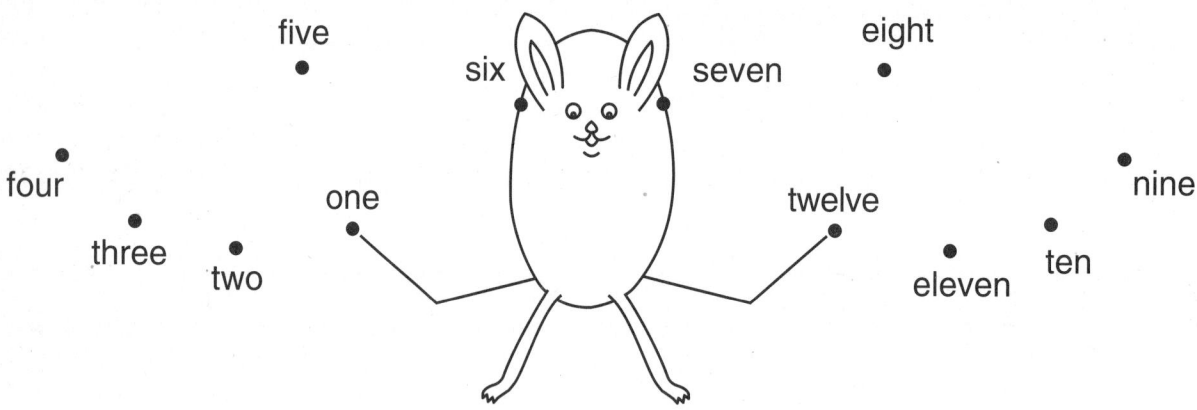

Name ________________________________

Cross Curricular
Activity 2

Fish and Birds

Connection to Reading
Science helps you understand what you read.

Read.

Animals are different.

Fish live in water.

Fish have scales.

Fish swim.

Birds live on land.

Birds have wings and feathers.

Birds fly.

Write.

How are birds different from fish?

____

____

Getting Milk

Connection to Math and Art

Science helps you put things in order
and write numbers.

Number these pictures in order.

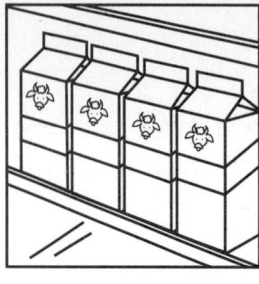
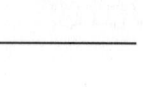

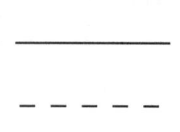

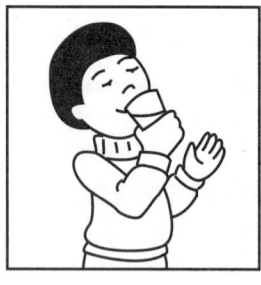
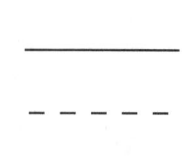

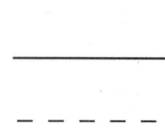

Water Animals

Connection to Art and Social Studies

Science helps you understand how animals are alike and different.

Circle the animals that live in water.

Make an X on each animal that lives on land.

Draw a water animal here.

Animal Homes

Connection to Language Arts
Science helps you understand words.

Draw lines to match the animals to their homes. Then write the name of the home next to its picture.

| cave   nest   pond |

_____
- - - - - - - - - - - - -
_____

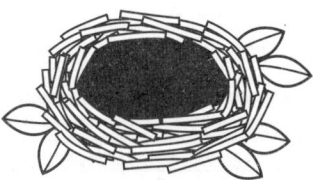

_____
- - - - - - - - - - - - -
_____

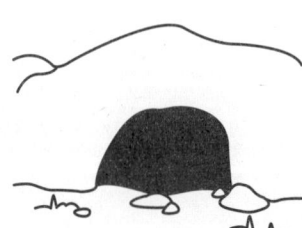

_____
- - - - - - - - - - - - -
_____

Camouflage

Connection to Language Arts and Art
Science helps you understand how animals behave.

Some animals and insects use their color to stay safe.

Find the hidden animals. Then color the rest of the picture to make them look hidden.

© Macmillan / McGraw-Hill

My Pet Book

My Pet Book

A Number of Shapes

Connection to Math and Language Arts
Science helps you tell how many.

What do you see in the sky?

Count and draw a line to match.

three clouds

one Sun

one Moon

four stars

Moon Trip

Connection to Reading and Language Arts
Science helps you ask questions.

You can take a vacation
to the Moon.

What three questions would
you ask before going?

_____

1. _____

_____ ?

2. _____

_____ ?

3. _____

_____ ?

Shape Change

Connection to Math and Reading
Science helps you order things.

Eli went on a long boat trip.

The first few days he did not see the Moon.

Then the Moon began to change shape.

Color the Moon that Eli saw **first**. Use a red crayon.

Color the Moon that Eli saw **next**. Use an orange crayon.

Color the Moon that Eli saw **last**. Use a yellow crayon.

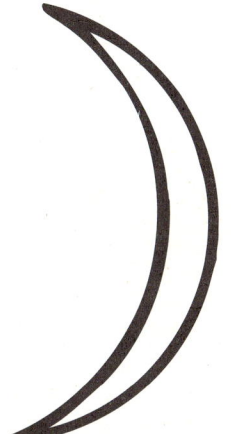
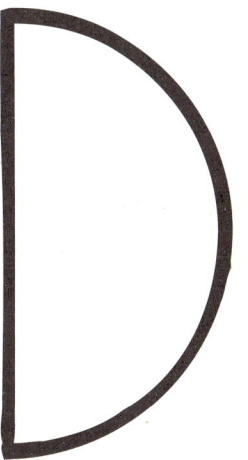
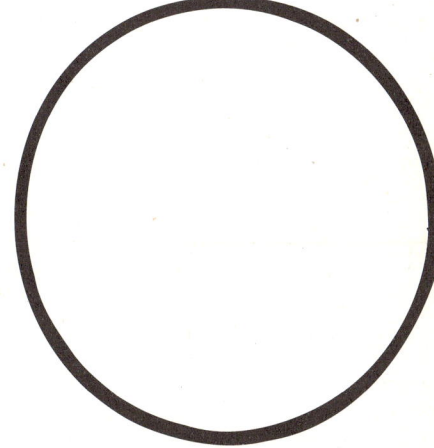

Weather Words

Connection to Language Arts
Science helps you understand weather words.

What is the weather in the picture?
Write a word from the box under its picture.

| sunny   rainy   snowy   foggy |

_____

_____

_____

_____

_____

_____

_____

_____

What to wear?

Connection to Math and Social Studies
Science helps you decide what to wear.

In the summer the weather is warm.
Circle the clothes you wear when it's warm.

In the winter the weather is cold in many places.
Place an X on each thing you wear when it's cold.

Name ______________________________

Cross Curricular
Culminating Activity

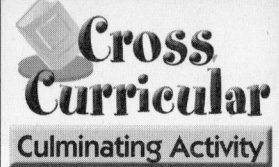

My Book of Stars

Year-Round Stars

Big Dipper

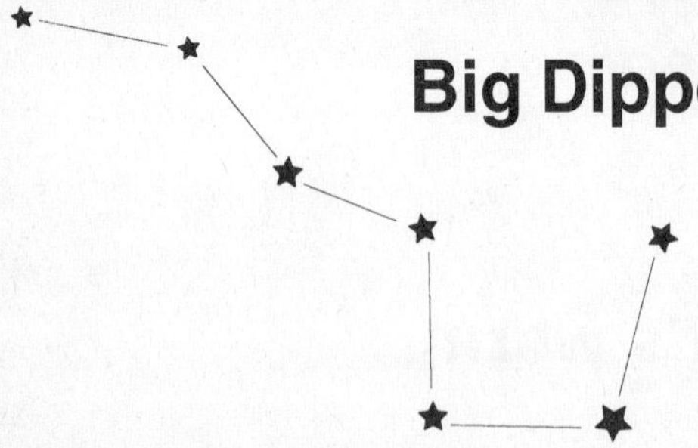

____

____

____

My Book of Stars

____

____

____

Summer Stars

Cygnus the Swan

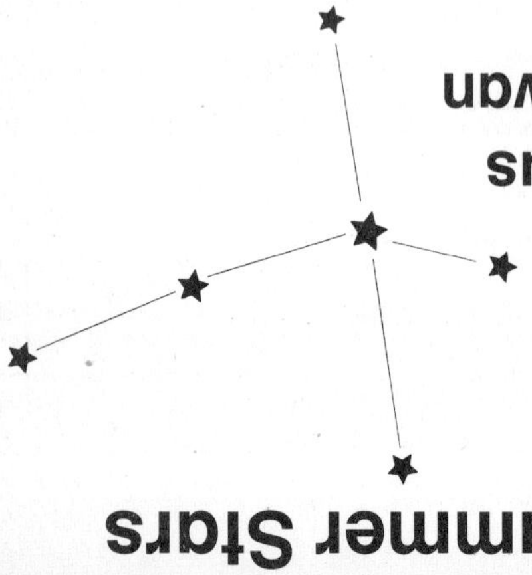

Winter Stars

Orion the Hunter

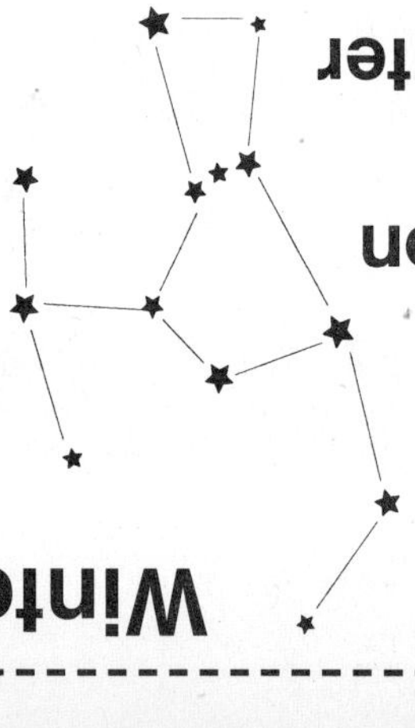

In the Soil

Connection to Art

Science helps you put things where they belong.

These things live in the soil.
Can you name them?

Draw them in the soil below.

Name ________________________________

Cross Curricular
Activity 2

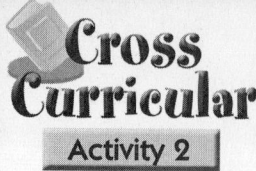

Water, Water

Connection to Social Studies and Art
Science helps you understand your planet.

Most of Earth is covered with water.
Color all the water blue.

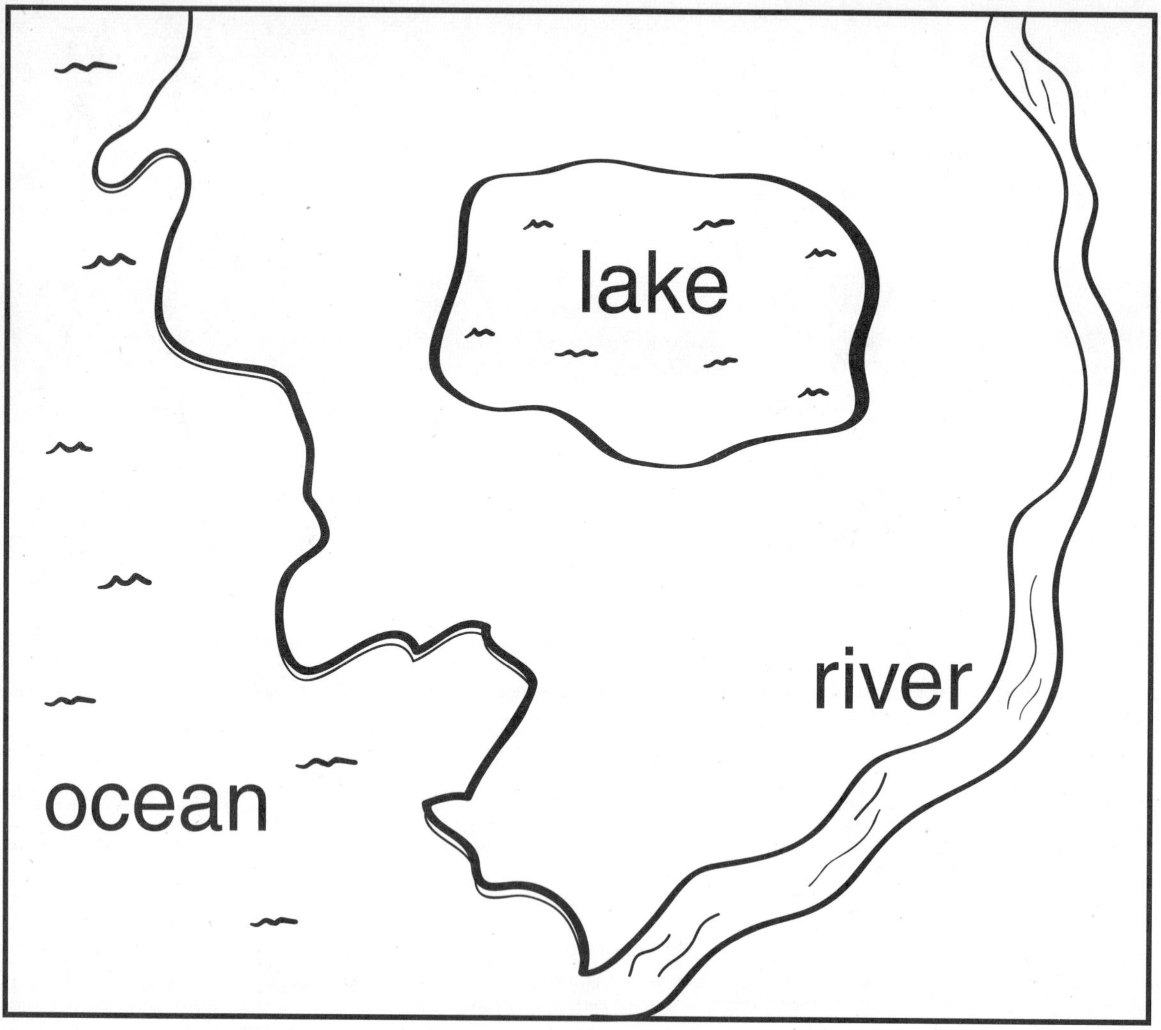
lake
river
ocean

Name ____________________

Cross Curricular
Activity 3

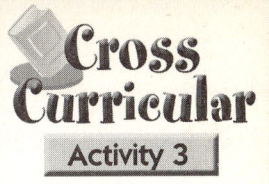

Rain to the Sea

Connection to Reading
Science helps you guess what will happen next.

Where do you think the rain will go?

Follow the stream with your finger.
Color the water blue.

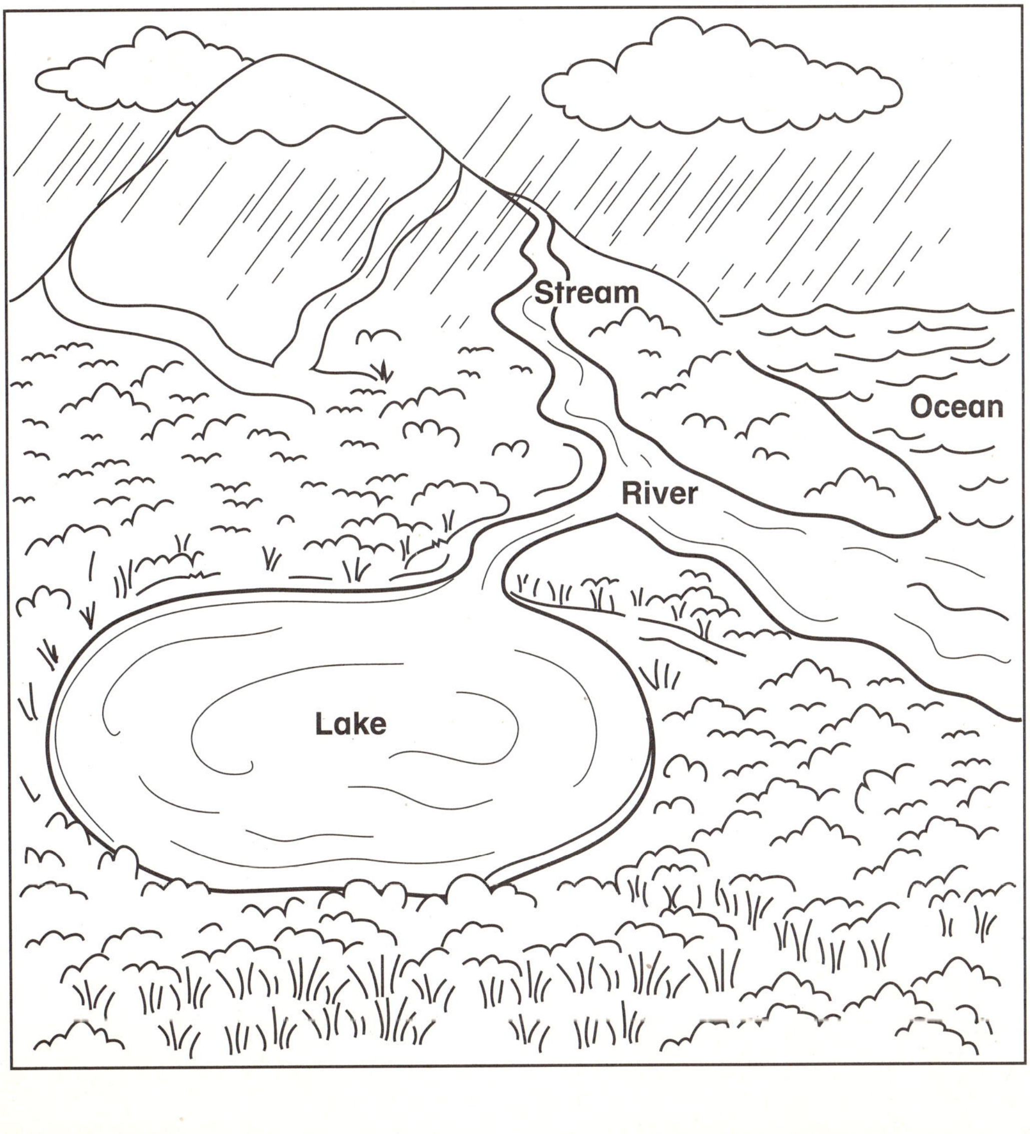

Stream
Ocean
River
Lake

The Air up There

Connection to Social Studies
Science helps you understand the world around you.

Circle the things that use air to move.

Stop Pollution

Connection to Social Studies
Science helps you take care of Earth.

Place an X on each thing that makes Earth dirty.

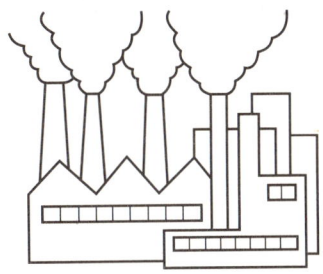

Play It Again

Connection to Art and Math
Science helps you learn how to recycle.

Draw lines to match.

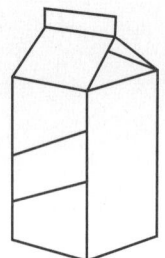

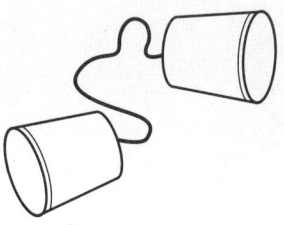

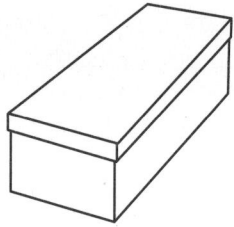

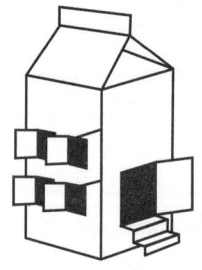

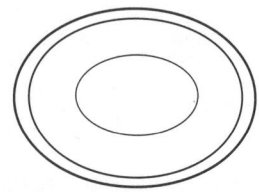

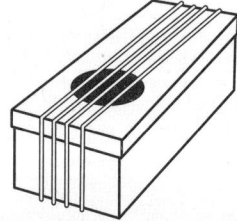

Cross Curricular
Culminating Activity

My Garden Book

My Garden Book

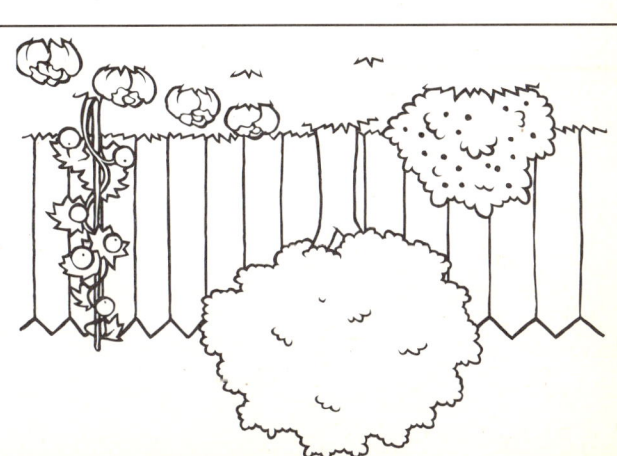

How heavy is it?

Connection to Math and Art
Science helps you predict how heavy things are.

Look at the toys.

Circle the lightest toy. Put an X on the heaviest toy.

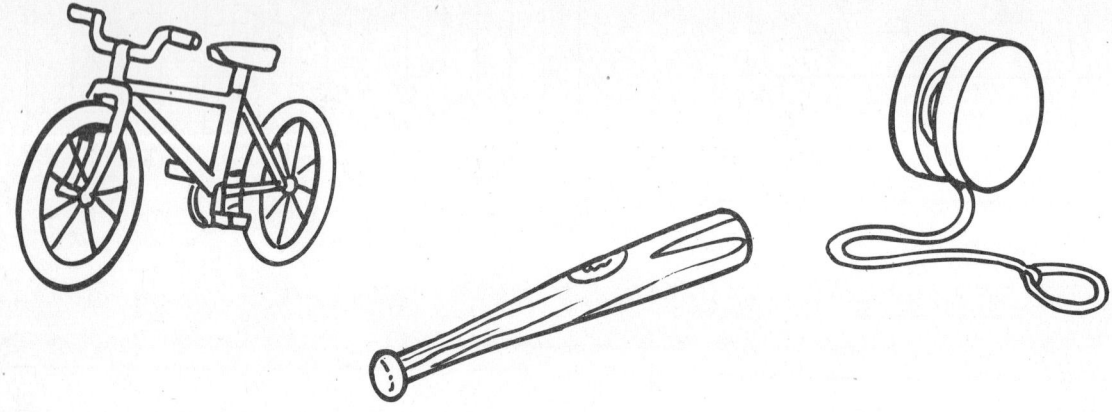

Draw an object that is heavy. Draw one that is light.

Lemonade Time

Connection to Math
Science helps you measure.

How much space does the lemonade take up?

Pour lemonade by coloring 12 squares in each cup.

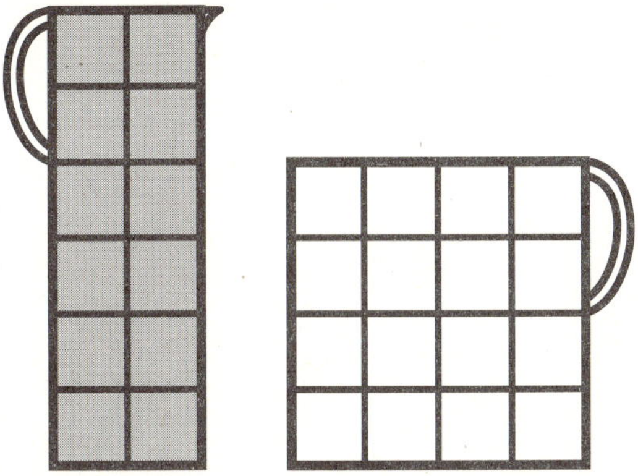

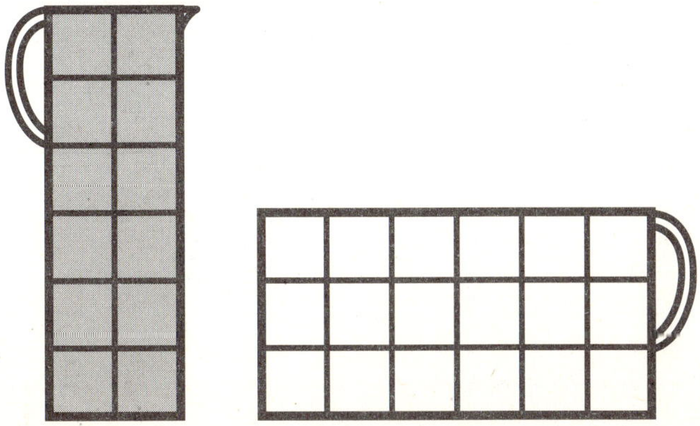

Lunch Time

Connection to Social Studies
Science helps you tell how things are alike and different.

Write S if the food is a solid.

Write L if the food is a liquid.

© Macmillan/McGraw-Hill

The Matter Song

Connection to Music
Science helps you write the words to a song.

The Matter Song

Solids and liquids are made of matter,
Made of matter, made of matter.
Solids and liquids are made of matter.
Gases are made of matter, too.

A solid has a shape and size,
Shape and size, shape and size.
A solid has a shape and size.
It keeps its shape and size.

A liquid has no shape of its own,
Shape of its own, shape of its own.
A liquid has no shape of its own.
It takes the shape of what it's in.

What do you know about gases?
Write something about gases.

Sink or Float

Connection to Language Arts
Science helps you use words to show what will happen next.

Some things float.

Some things sink.

Circle the things that float.

Make an X on each thing that sinks.

How hot?

Connection to Math
Science helps you measure and count.

Look at the thermometers. Read the temperatures.

Write how hot it is.

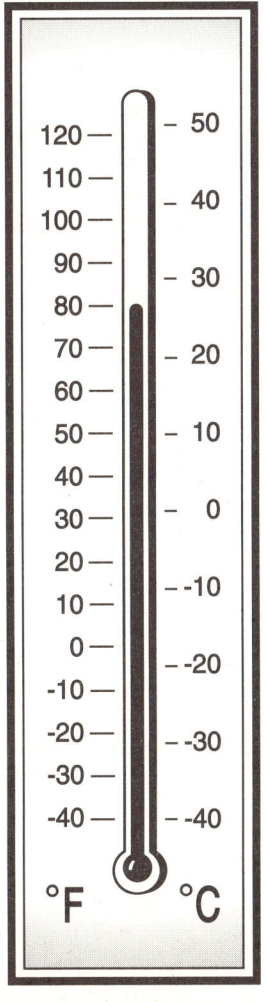

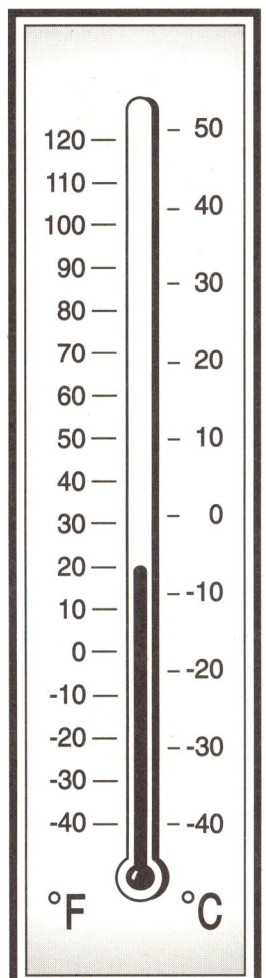

_____

_ _ _ _ _ _ _ _ _ _ _

_____ °F

_____

_ _ _ _ _ _ _ _ _ _ _

_____ °F

My Snow Book

_____

- -

_____

My
Snow
Book

_____

- -

_____

_____

- -

_____

_____

- -

_____

Name ______________________

Cross Curricular
Activity 1

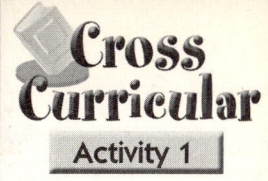

Push and Pull

Connection to Reading
Science helps you guess what will happen next.

James is going to move these things.

Write **push** next to the things that will move when he pushes them.

Write **pull** next to the things that will move when he pulls.

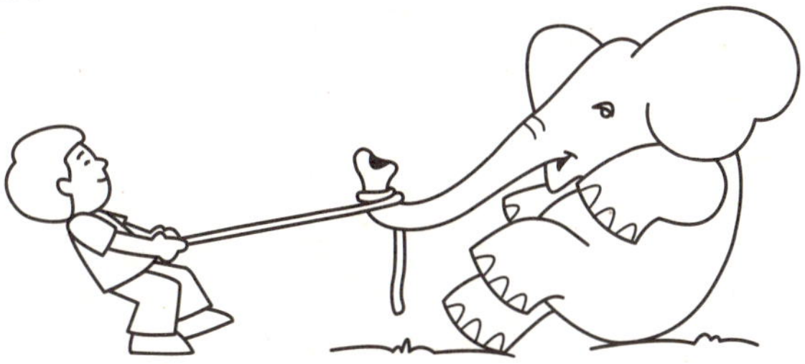

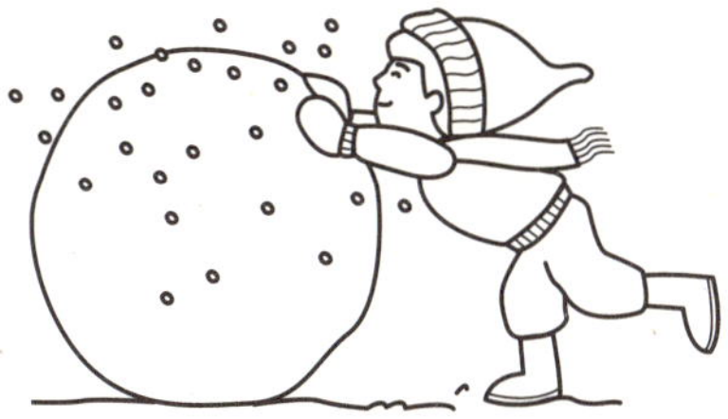

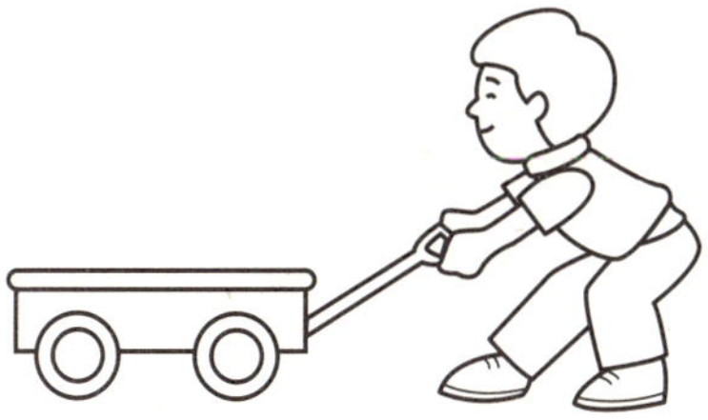

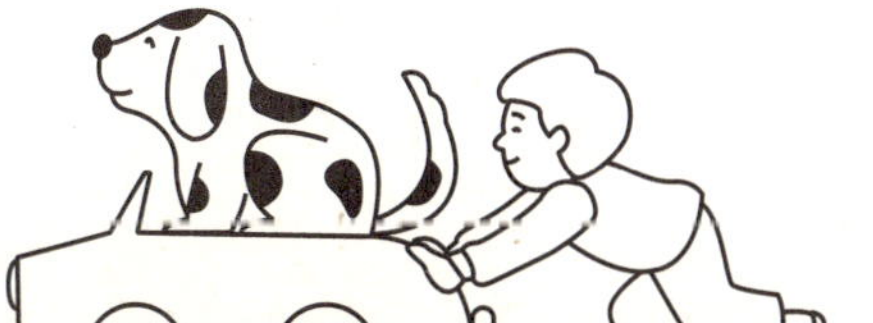

Where is it?

Connection to Language Arts and Reading
Science helps you communicate.

Write a word from the box to complete each sentence.

| in   across   under   on   over |

1. The boy is _____ the swing.

2. The cat ran _____ the grass.

3. The girl jumped _____ the rock.

4. The duck is _____ the pond.

5. The dog sat _____ the tree.

Name ______________________________

Cross Curricular
Activity 3

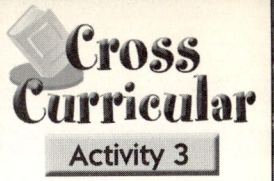

From Here to There

Connection to Language Arts and Math
Science helps you understand position words.

The monkey is on the move.

Follow the path that takes it **under, over, around,** and **through.**

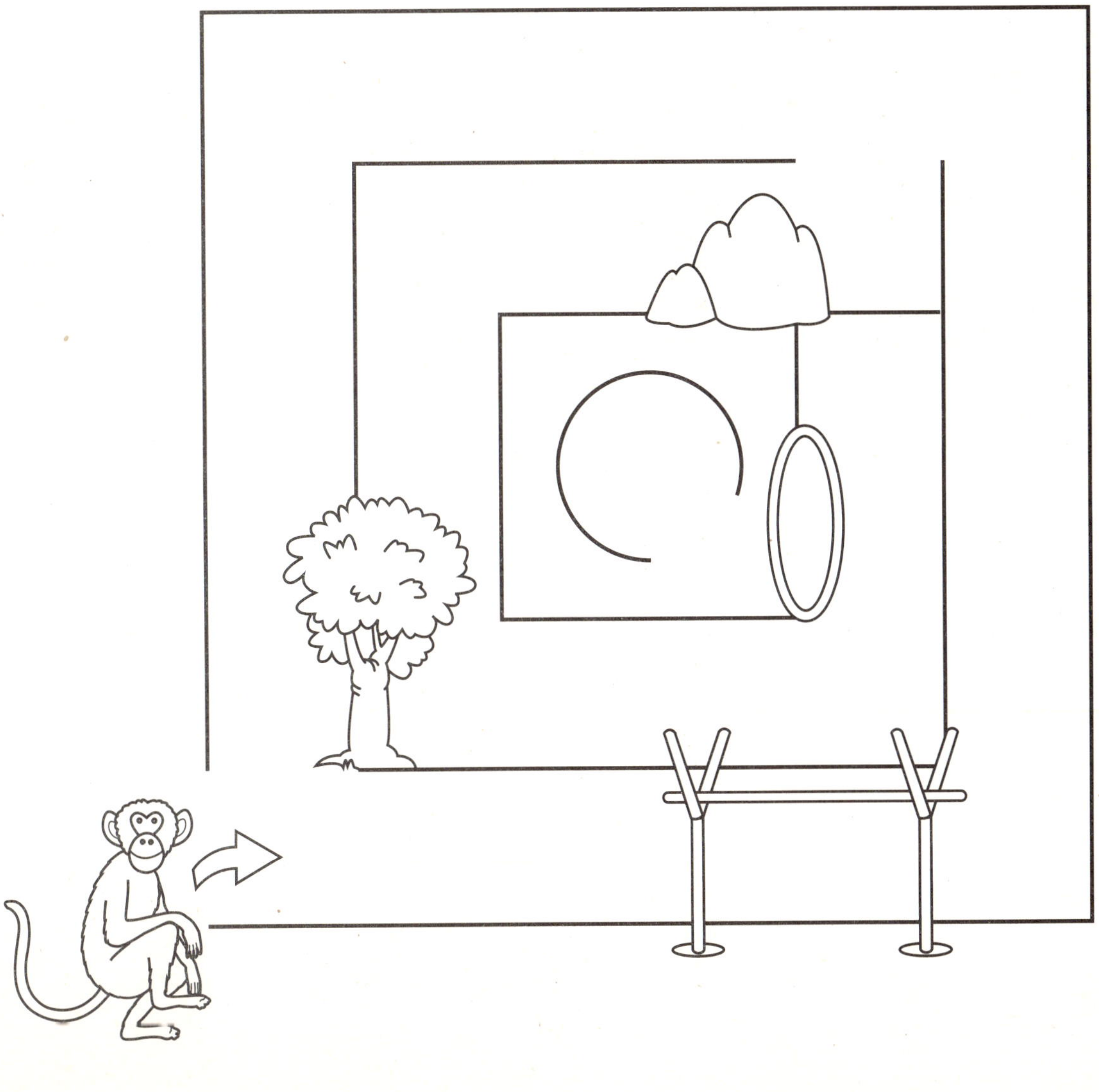

Move it!

Connection to Reading

Science helps you understand how some things are alike.

Pins, nails, and paper clips are alike.
They are all metal.
A magnet can pick up metal.

Circle the things that a magnet can pick up.

 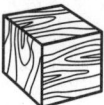  

  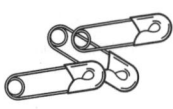

What moves it?

Connection to Reading

Science helps you guess how some things move.

How will these things move?

Choose a word for each sentence.

Write it.

| fly | turn | pull | hit |

_____

The magnet will _____ the nails up.

_____

The girl will _____ the ball over the fence.

_____

The children will _____ the jump rope.

_____

The pilot will _____ the airplane.

Missing Pieces

Connection to Art

Science helps you finish pictures with missing pieces.

Sound happens because something moves.

Read what is missing. Draw it in the picture.

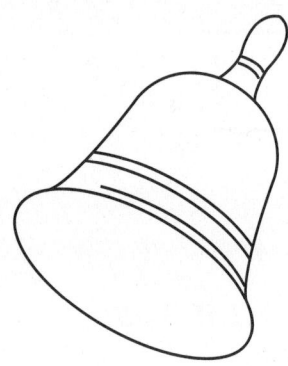

missing ringer

missing strings

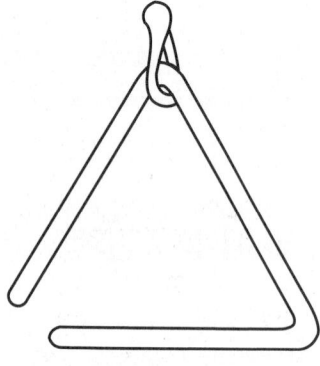

missing stick

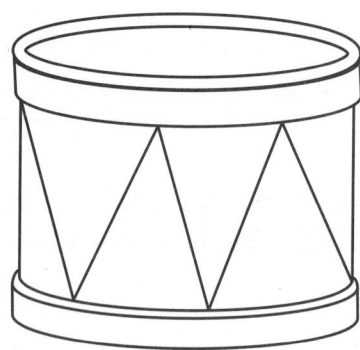

missing drumsticks

My Playground Book

My Playground Book

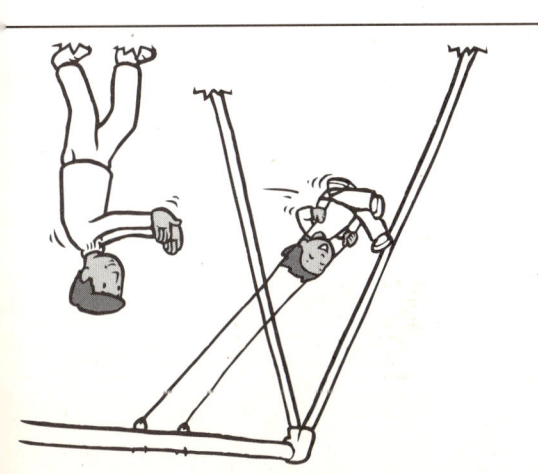

Project Theme

A Walk in the Woods, pages 1–7

Concepts
- Apply knowledge of plants through art and writing.
- Show understanding that living things change.
- Make a graph and compare numbers of objects.
- Use vocabulary words in new situations.
- Use pictures to gather information.

Overview In **A Walk in the Woods**, children use language arts and reading skills to write, spell, and apply vocabulary words; math skills to make a graph and compare numbers of objects; and social studies skills to predict changes in seasons. The unit culminates with a writing activity about plants.

Getting Started Introduce **A Walk in the Woods** as the unit theme. Ask children if they have ever taken a walk in the woods. If so, what did they see, hear, touch, and smell? If children have not been in the woods, ask if they have taken a walk in a park where a lot of trees grow. You may wish to bring in photographs of trees and woodlands to use as a prompt for the discussion.

A Walk in the Woods Planning Guide

Grade 1 Unit A: Plants Are Living Things	Activity	Resources	Related Subject	Macmillan/McGraw-Hill Programs
Chapter 2: A Look at Plants **Lesson 4:** Plants Have Parts	**Activity 1:** Plants Have Parts, p. 1		Language Arts	*McGraw-Hill Language Arts,* Grade 1, pp. 97–102
Chapter 2: A Look at Plants **Lesson 6:** Stems and Leaves	**Activity 2:** All Fall Down, p. 2		Math	*McGraw-Hill Mathematics,* Grade 1, pp. 93–94
Chapter 2: A Look at Plants **Lesson 7:** Seeds	**Activity 3:** A Walk in the Woods, p. 3		Reading and Social Studies	*McGraw-Hill Reading,* Grade 1, Book 3, *Sam's Song* *My World: Adventures in Time and Place,* Grade 1, pp. 110–113
Chapter 2: A Look at Plants **Lesson 7:** Seeds	**Activity 4:** What plant is it? p. 4		Language Arts	*McGraw-Hill Language Arts,* Grade 1, pp. 39–42
Chapter 2: A Look at Plants **Lesson 8:** Plants Grow and Change	**Activity 5:** Plants Graph, p. 5		Math	*McGraw-Hill Mathematics,* Grade 1, pp. 257–258
Chapter 2: A Look at Plants **Lesson 8:** Plants Grow and Change	**Activity 6:** A Tree Changes, p. 6	*Have You Seen Trees?*, Joanne Oppenheim	Social Studies	*My World: Adventures in Time and Place,* Grade 1, pp. 110–113
Chapters 1 and 2: All About Living Things; A Look at Plants **All Lessons**	**Culminating Activity:** My Book of Plants, p. 7	*Have You Seen Trees?*, Joanne Oppenheim	Reading and Language Arts	*McGraw-Hill Reading,* Book 4, Johnny Appleseed *McGraw-Hill Language Arts,* Grade 1, pp. 379–380

Scoring Rubric for Integration Activities

Score	Criteria
4	Accomplished all of the activity's objectives.
3	Accomplished more than half of the activity's objectives.
2	Accomplished less than half of the activity's objectives.
1	Made little or no progress toward accomplishing the activity's objectives.

Activity 1

Connection to Language Arts

Plants Have Parts, page 1

Objective: Identify illustrations that correspond to written vocabulary words.

Introduce: Ask children to imagine they are walking in the woods with a friend. The friend wants to know about plants. Ask children to name words that describe the parts of a plant. List the words on the board.

Teach: Have children draw a line from each word to the picture that matches it.

Close: Ask, *Which word was easiest to match with its picture. Why? Which word was the hardest?* Let children share their work.

Assessment: Children should be able to match pictures of plant parts to the corresponding words.

Modification: Children may need to say each word aloud.

Answers: Lines should connect each word with the correct picture.

Activity 2

Connection to Math

All Fall Down, page 2

Objective: Count and compare numbers of leaves.

Introduce: Review numbers 1 through 10.

Teach: Ask volunteers to count the numbers of leaves. Write each answer on the board.

Close: Ask which has more leaves–the tree or the ground? On the board, write a subtraction sentence to show the difference.

Assessment: Children should count and compare the number of leaves on the ground and on the tree.

Modification: Invite children to draw a tree with a different number of leaves and challenge partners to count and compare the flowers they drew.

Answers: 4; 5; 9; the tree has one more leaf than the ground.

Activity 3

Connection to Reading and Social Studies

A Walk in the Woods, page 3

Objective: Use information from the poem to put illustrations in correct sequence.

Introduce: Ask children if they have seen squirrels in the woods or a park. Invite them to describe how squirrels behave.

Teach: Invite volunteers to sound out the poem and read it aloud. Then have children put the illustrations in order.

Close: Let volunteers share their answers. Discuss how each line of the poem matches each picture.

Assessment: Children should sequence the pictures with the appropriate numbers.

Modification: If children have difficulty reading the poem, write the lines on the board and demonstrate blending.

Answers: Pictures should be numbered according to the order of the poem.

Activity 4

Connection to Language Arts

What plant is it?, page 4

Objective: Use a code to spell science vocabulary.

Introduce: Ask children if they have ever used a secret code. Use the activity sheet to explain how a code works.

Teach: Lead the class in reading the page. Explain that they will use the code to match each number with a letter.

Close: Have volunteers tell how they used the code to find the words. Then let them suggest other plant words and find any letters from those words in the code.

Assessment: Children should use the code to spell science words.

Modification: Some children may not use the code. Ask them to check their answers by using the code.

Answers: warm, water, tree, lemon.

Activity 5

Connection to Math

Plants Graph, page 5

Objective: Make a graph to show the number of plants in a picture.

Introduce: Ask children to name as many plants as they can.

Teach: Point to the bar graph for the trees and the one for the flowers. Explain to children that they have to color in the graphs to show the number of trees and the number of flowers in the picture.

Close: Ask, *What does the graph show?* Have children use their graphs to answer questions such as, *Are there more trees or flowers?*

Assessment: Children should color a box in the appropriate graph for each plant in the picture.

Modification: Children may use stickers or cut-outs to make the graph.

Answers: The tree graph should be shaded up to the number 3. The flowers graph should be shaded up to the number 5.

Activity 6

Connection to Social Studies

A Tree Changes, page 6

Objective: Predict how a tree changes with the season.

Introduce: Discuss what changes occur during each season. Consider questions such as *How do plants change with the seasons?* Tell children that science helps you understand such patterns in nature.

Teach: Have children use crayons for the activity. Have children look at the drawing of a tree in Fall. Ask, *What will the same tree look like in Winter?*

Close: Display children's drawings. Point to the drawings for Winter and ask, *How do we know that the tree is in a place where winters are cold?*

Answer: The tree has lost its leaves.

Assessment: Children should be able to use their knowledge of the seasons to predict what happens to the tree's leaves.

Modification: Sight-impaired children might describe how fallen leaves feel and sound as they walk in them.

Answers: Winter drawing should show bare trees. Spring drawings should show flowers and/or small leaves. Summer drawing should show full-grown leaves.

Culminating Activity

Connection to Reading and Language Arts

My Book of Plants, page 7

Objectives: Use pictures to gather information to use in a story about plants.

Introduce: Explain that there are books called field guides that contain the names of plants. If possible, show one to the class.

Teach: Help children fold the page to make a 4-page book. Lead them in studying the pictures in the book. Children then use the pictures to write a story about what they would expect to see as they walk in the woods.

Close: Let children share their stories and display them in the classroom.

Enrichment: Challenge children to make a field guide to plants in your area.

Assessment: Children should use the pictures in the book as a resource as they write their own story.

Modification: Children may include drawings from the field guide in their story.

Answer: Children's stories will vary, but should discuss seeds, leaves, and fruit.

Project Theme

Animals Around Us, pages 8–14

Concepts
- Group animals by common characteristics.
- Make comparisons based on a passage.
- Use vocabulary words in writing.
- Show understanding that living things change.
- Identify a sequence of events.
- Identify appropriate animal habitats.

Overview In **Animals Around Us**, children use language arts and reading skills to write, spell, and apply science vocabulary words; math skills to demonstrate understanding of sequencing events; art skills to explore environmental adaptation; and social studies skills to explore change in living things. The unit culminates with a book that children use as a resource for a writing activity about animals.

Getting Started Introduce **Animals Around Us** as the unit theme. Ask children what animals they see every day. Children may suggest pets (dogs, cats, goldfish), wild animals (squirrels, snakes, and birds), and insects (butterflies and flies). Encourage children to name their favorite animals. If you wish, each child can choose a mascot to use during the unit. Children can decorate their desks or cubbyholes with art relating to their mascots.

Animals Around Us Planning Guide

Grade 1 Unit B: Animals Are Living Things	Activity	Related Subject	Macmillan/McGraw-Hill Programs
Chapter 3: A Look at Animals **Lesson 2:** Mammals	**Activity 1:** I Am a Mammal, p. 8	Math and Language Arts	*McGraw-Hill Mathematics*, Grade 1 pp. 19–20 *McGraw-Hill Language Arts*, Grade 1, pp. 203–214
Chapter 3: A Look at Animals **Lesson 3:** More Animals Groups	**Activity 2:** Fish and Birds, p. 9	Reading	*McGraw-Hill Reading*, Grade 1, Book 5, The Story of Blue Bird
Chapter 4: How Animals Meet Their Needs **Lesson 5:** Getting Food	**Activity 3:** Getting Milk, p. 10	Math and Art	*McGraw-Hill Mathematics*, Grade 1, pp. 53–54
Chapter 4: How Animals Meet Their Needs **Lesson 6:** Where Animals Live	**Activity 4:** Water Animals, p. 11	Art and Social Studies	*My World: Adventures in Time and Place*, Grade 1, pp. 120–127
Chapter 4: How Animals Meet Their Needs **Lesson 6:** Where Animals Live	**Activity 5:** Animal Homes, p. 12	Language Arts	*McGraw-Hill Language Arts*, Grade 1, pp. 203–214
Chapter 4: How Animals Meet Their Needs **Lesson 7:** Staying Safe	**Activity 6:** Camouflage, p. 13	Language Arts and Art	*McGraw-Hill Language Arts*, Grade 1, pp. 203–214
Chapters 3–4: A Look at Animals; How Animals Meet Their Needs **All Lessons**	**Culminating Activity:** My Book of Pets, p. 14	Reading, Language Arts, and Social Studies	*McGraw-Hill Reading*, Book 2, A Vet *McGraw-Hill Language Arts*, Grade 1, pp. 203–214 *My World: Adventures in Time and Place*, Grade 1, pp. 120–127

Scoring Rubric for Integration Activities	
Score	**Criteria**
4	Accomplished all of the activity's objectives.
3	Accomplished more than half of the activity's objectives.
2	Accomplished less than half of the activity's objectives.
1	Made little or no progress toward accomplishing the activity's objectives.

Activity 1

Connection to Math and Language Arts

I Am a Mammal, page 8

Objective: Follow number words to connect the dots.

Introduce: Ask children what animals they know are mammals. Remind them that humans are mammals.

Teach: Lead the class in reading the instructions. Let children work at their own pace to connect the dots.

Close: Ask children to name the mammal that they drew.

Assessment: Children should correctly connect the dots to make an outline of a bat.

Modification: Children can write the numerals next to the number words.

Activity 2

Connection to Reading

Fish and Birds, page 9

Objective: Understand how birds and fish are different.

Introduce: Display a photo of a fish and a photo of a bird, perhaps a parrotfish and a parrot. Invite volunteers to say how they are alike and different.

Teach: Lead children in reading the page aloud. Have children write a few words showing one difference between birds and fish.

Close: Invite volunteers to share their answers.

Assessment: Children should show understanding of the concept with a few words.

Modification: Children who have difficulty writing can explain orally to a partner how fish and birds are different.

Activity 3

Connection to Math and Art

Getting Milk, page 10

Objective: Understand the order of events.

Introduce: Show the children the four pictures.

Teach: Help the children read the directions. Have children number the pictures in order of the events.

Close: Invite volunteers to share their answers.

Assessment: Children should show understanding of the concept by correctly numbering the pictures.

Modification: Visually impaired children can listen to descriptions of the pictures and order them accordingly

Answers: 1 (cow eating grass), 2 (cow being milked), 3 (milk cartons in store), 4 (child drinking milk)

Activity 4

Connection to Art and Social Studies

Water Animals, page 11

Objective: Compare different types of animals based on their natural habitats.

Introduce: Ask *What animals live in the water? What animals live on land?* Write the animal names in two lists on the board.

Teach: Lead the class in naming the animals on the activity sheet. For each animal, ask, *Does (a zebra) live in the water? Does (a zebra) live on land?*

Close: Display children's drawings. Invite discussion of the animals drawn.

Assessment: Children should correctly identify land and water animals.

Modification: Early finishers can draw a water animal.

Answers: (circle) whale, dolphin, fish, octopus. (cross) mouse, zebra, tiger.

Activity 5
Connection to Language Arts

Animal Homes, page 12

Objective: Match animals to their homes.

Introduce: Ask children if they have observed animal homes in their neighborhood. Have children list as many animal homes as possible.

Teach: Invite children to complete the activity.

Close: Discuss how animals build their homes.

Assessment: Children should match each animal to its home and write the correct home next to its picture.

Modification: Visually impaired children can spell out the correct answers.

Answers: fish/pond, bird/nest, bat/cave

Activity 6
Connection to Language Arts and Art

Camouflage, page 13

Objective: Understand how animals stay safe by blending into their surrounding environment.

Introduce: Ask children if they have noticed animals that blend into the grass or trees. Show photographs of animals in nature that are camouflaged.

Teach: Invite children to find the animals in the woods. Children can color the animals, then color the background to make them "hide."

Close: Discuss how each animal is camouflaged. Ask *Why is camouflage important to animals?*

Assessment: Children should have the animals look hidden.

Modification: Visually impaired children can describe how various animals blend into a natural background.

Answers: Students should identify and color the lizard, grasshopper, bird, deer, and fox.

Culminating Activity

Connection to Reading, Language Arts, and Social Studies

My Pet Book, page 14

Objective: Use pictures to gather information to write about pets.

Introduce: Read a story about pets, then use it as a reference for writing a story.

Teach: Help children cut the page and then fold it to make a 4-page book. Tell children to use the book you read to gather information for a story about how to take care of a pet.

Close: As children share their stories, make a list of the different pets they wrote about.

Enrichment: Make a graph of the pets in children's stories and find the most common class pet.

Assessment: Children should use the hamster pictures as a resource as they write their own story about pets.

Modification: Children may write about a pet at home or a classroom pet, or they can pretend they have a pet.

Answer: Children's stories will vary.

Project Theme

Going on Vacation, pages 15–20

Concepts
- Identify and count objects in the sky.
- Ask questions about Moon travel.
- Order phases of the Moon.
- Label illustrations of weather words.
- Classify appropriate clothing for winter and summer.
- Use pictures to gather information.

Overview In **Going on Vacation,** children use the theme of a vacation as a context for exploring weather and the stars. Children use social studies skills to infer about the seasons; math skills to order numbers; language arts skills to use science vocabulary; and reading skills to order events and identify questions for inquiry. The unit culminates with a picture book for children to write a story about a trip to the stars.

Getting Started The theme for this unit is **Going on Vacation.** Ask where the children spend their vacations. Discuss which vacation time—summer, winter, spring, or fall—they enjoy the most and why.

Going on Vacation Planning Guide

Grade 1 Unit C: The Sky and Weather	Activity	Resources	Related Subject	Macmillan/ McGraw-Hill Programs	Materials
Chapter 5: The Sky **Lesson 1:** The Sun	**Activity 1:** A Number of Shapes, p. 15		Math and Language Arts	*McGraw-Hill Mathematics*, Grade 1 Chapter 1, *McGraw-Hill Language Arts*, Grade 1, pp. 91–94	
Chapter 5: The Sky **Lesson 2:** The Moon and Stars	**Activity 2:** Moon Trip, p. 16	*Roaring Rockets*, Tony Mitton and Ant Parker	Reading and Language Arts	*McGraw-Hill Reading*, Grade 1, pp. 228–289 *McGraw-Hill Language Arts*, Grade 1, pp. 3–18	
Chapter 5: The Sky **Lesson 2:** The Moon and Stars	**Activity 3:** Shape Change, p. 17	*The Moon Seems to Change*, Franklyn M. Branley	Math and Reading	*McGraw-Hill Mathematics*, Chapter 1 *McGraw-Hill Reading*, Grade 1, pp. 121–139	Calendar or reference materials that identify sequence of Moon phases
Chapter 6: Weather and Seasons **Lesson 5:** Weather Changes	**Activity 4:** Weather Words, p. 18	*How's the Weather?*, Melvin and Gilda Berger; *Sun Snow Stars Sky*, Catherine and Lawrence Anholt	Language Arts	*McGraw-Hill Language Arts*, Grade 1, pp. 91–99	
Chapter 6: Weather and Seasons **Lesson 5:** Weather Changes	**Activity 5:** What to wear?, p. 19	*Sun Snow Stars Sky*, Catherine and Lawrence Anholt	Math and Social Studies	*McGraw-Hill Mathematics*, Grade 1 Chapter 11 *My World: Adventures in Time and Place*, Grade 1, pp. 137–141	
Chapter 5: The Sky **All Lessons**	**Culminating Activity:** My Book of Stars, p. 20		Reading and Language Arts	*McGraw-Hill Reading*, Grade 1, pp. 121–139 *McGraw-Hill Language Arts*, Grade 1, pp. 91–99	

Scoring Rubric for Integration Activities	
Score	**Criteria**
4	Accomplished all of the activity's objectives.
3	Accomplished more than half of the activity's objectives.
2	Accomplished less than half of the activity's objectives.
1	Made little or no progress toward accomplishing the activity's objectives.

Activity 1

Connection to Math and Language Arts

A Number of Shapes, page 15

Objective: Match sky and weather words with pictures.

Teach: Read the words with children. Ask them which picture goes with each word.

Close: Ask the children, *Can you see the objects shown during the day or night?* It is possible to see clouds at night, and the Moon and the stars early in the morning.

Assessment: Children should correctly match each picture with the words.

Activity 2

Connection to Reading and Language Arts

Moon Trip, page 16

Objective: Create questions about taking a trip to the Moon.

Introduce: Read aloud a story about outer space or space travel. Show pictures of the Moon, astronauts, and rockets.

Teach: Invite children to do the activity. Encourage them to write questions about any aspect of a Moon trip.

Close: Let volunteers read their questions. Then use books or the Internet to find answers to the questions children recorded.

Assessment: Children should be able to create three questions about Moon travel.

Modification: Some children might benefit from looking at the book you read while they record their questions.

Answers: Questions will vary.

Activity 3

Connection to Math and Reading

Shape Change, page 17

Objectives: Sequence the changing shape of the Moon.

Introduce: Use the class schedule to demonstrate the terms first, next, and last.

Teach: Gather children around a calendar that identifies in pictures the days on which there is a New Moon, Full Moon, and so on. If a calendar is not available, locate such pictures in an encyclopedia or other reference material. Find the New Moon and explain that Eli and his family start a boat trip on that day, are away for many days, and return when the Moon is full. Read the page aloud and have children complete the activity.

Close: Find out the current shape of the Moon. Ask children if they can predict how it will change in the coming weeks.

Assessment: Children should be able to identify the sequence of the changing shapes of the Moon.

Modification: Some children might need to sketch the phases of the Moon in order before applying the terminology.

Answers: Crescent is first (red); First Quarter is next (orange); Full is last (yellow)

Activity 4

Connection to Language Arts

Weather Words, page 18

Objective: Match science vocabulary words to weather scenes.

Introduce: Share a printed weather report from a newspaper and discuss the words used to describe the weather. Challenge children to brainstorm a list of words that describe weather. Then have children define each word.

Teach: Read the activity directions. Be sure children understand they are to use each word once.

Close: Ask volunteers to share their answers. Have them describe the word *foggy*.

Enrichment: Children find magazine pictures that show the other weather words the class listed.

Assessment: Children use what they know about weather to help them understand new words.

Modification: Some children may need help reading the words in the box.

Answers: Top left: snowy; top right: foggy; bottom left: rainy; bottom right: sunny.

Activity 5

Connection to Math and Social Studies

What to wear?, page 19

Objectives: Identify appropriate clothing for different seasons.

Introduce: Discuss how you decide what clothes you need on vacation. Help children realize that they must consider the season, the weather, and what their needs will be during the vacation.

Teach: Have a volunteer read the directions aloud. Point out the first thing they must do is decide what one wears when it is cold and snowy, or when it is sunny and hot.

Close: Have volunteers share their answers. Then have children consider a tee shirt. Discuss how a person might need a tee shirt in both seasons.

Enrichment: Ask, *What clothes might you need if you were going on a camping trip in the woods during the spring?* Have children draw a picture of the clothes and explain the weather they expect.

Assessment: Children should identify clothes for going to the mountains in winter and those for those going to the beach.

Modification: You may wish to bring in clothing for children to sort by season.

Answer: (circle): shorts, beach slippers, sun hat; (X): gloves, scarf, boots, snowsuit, wool hat

Culminating Activity

Connection to Reading and Language Arts

My Book of Stars, page 20

Objective: Use pictures to gather information to write about the stars.

Introduce: Explain that there are field guides to the planets and stars. If possible, show one to the class.

Teach: Help children cut the page and fold it to make a 4-page book. Study the titles and pictures in the book. Children then use the titles and pictures to write a story about what they might expect to see if they took a trip in outer space. Encourage children to draw pictures of Orion the Hunter and Cygnus the Swan around the constellations.

Close: Let children share their stories and display them in the classroom.

Assessment: Children should use the pictures and titles of a book on outer space as a resource for writing their own stories.

Modification: Children may include drawings from the field guide in their stories.

Answer: Children's stories will vary.

Project Theme

A Living Planet, pages 21–27

Concepts
- Recognize and draw animals and parts of plants that live in the soil.
- Identify bodies of water on a map.
- Understand the path of water in a river system.
- Classify things that fly.
- Identify damaging forms of pollution.
- Recycle and create art projects from discarded items.

Overview The theme **A Living Planet** allows children to explore plant and animal life in a habitat. Children use art skills to draw life in the soil; language arts skills to apply vocabulary words; social studies skills to read maps; and reading skills to comprehend categories. The unit culminates with a book that children use about food sources in a garden.

Getting Started Introduce **A Living Planet** as the unit theme. Let children share what they know about plants and soil. Ask, *Do you have plants? Have you been to a park? What lives in the soil?* Impress upon children that people grow plants and take care of them. Help children to see that a habitat is where plants and animals live.

A Living Planet Guide

Grade 1 Unit D: Caring for Earth	Activity	Resources	Related Subject	Macmillan/McGraw-Hill Programs
Chapter 7: Earth's Resources **Lesson 2:** Soil	**Activity 1:** In the Soil, p. 21		Art	
Chapter 7: Earth's Resources **Lesson 3:** Water	**Activity 2:** Water, Water, p. 22		Social Studies and Art	*My World: Adventures in Time and Place*, Grade 1, pp. 104–124
Chapter 7: Earth's Resources **Lesson 3:** Water	**Activity 3:** Rain to the Sea, p. 23		Reading	*McGraw-Hill Reading, Day by Day* Grade 1, pp. 110–120
Chapter 7: Earth's Resources **Lesson 4:** Air	**Activity 4:** The Air up There, p. 24		Social Studies	*My World: Adventures in Time and Place*, Grade 1, pp. 104–124
Chapter 8: Taking Care of Earth **Lesson 6:** Pollution	**Activity 5:** Stop Pollution, p. 25		Social Studies	*My World: Adventures in Time and Place*, Grade 1, pp. 104–124
Chapter 8: Taking Care of Earth **Lesson 7:** Caring for Earth's Resources	**Activity 6:** Play It Again, p. 26		Art and Math	*McGraw-Hill Mathematics*, Grade 1, pp. 445–453
Chapters 7 and 8: Earth's Resources; Taking Care of Earth **All Lessons**	**Culminating Activity:** My Garden Book, p. 27		Reading, Social Studies, and Language Arts	*McGraw-Hill Reading*, Grade 1, pp. 49–79 *My World: Adventures in Time and Place*, Grade 1, pp. 104–124 *McGraw-Hill Language Arts*, Grade 1, pp. 117–138

Scoring Rubric for Integration Activities	
Score	**Criteria**
4	Accomplished all of the activity's objectives.
3	Accomplished more than half of the activity's objectives.
2	Accomplished less than half of the activity's objectives.
1	Made little or no progress toward accomplishing the activity's objectives.

Activity 1

Connection to Art

In the Soil, page 21

Objective: Identify animals and parts of plants that live in soil.

Introduce: Ask children if they have ever helped plant a garden. Discuss the animals they saw in the soil. Ask, *What part of the plant do you put underground?*

Teach: Ask volunteers to name the living things (ant, worm, mole, plant, tree) in the pictures, and then allow children to work independently to make their drawings.

Close: Display children's drawings. Discuss the importance of earthworms and other soil animals.

Modification: Early finishers can draw some animals that live above ground.

Activity 2

Connection to Social Studies and Art

Water, Water, page 22

Objective: Identify bodies of water on a map.

Introduce: Remind children that all living things depend on water. Explain that Earth is full of living things because of the plentiful supply of water and sunlight.

Teach: Guide children in reading aloud the labels of the bodies of water on the map. Then have children color the bodies of water.

Close: Use a globe to show children how much of Earth is covered with water.

Modification: Children can color the landforms on their maps green or yellow.

Activity 3

Connection to Reading

Rain to the Sea, page 23

Objective: Understand the path of water in a river system.

Introduce: Ask children to describe where a leaf goes if it falls into a stream.

Teach: Review the names of the bodies of water on the activity page. Allow children to complete the activity.

Close: Invite children to speculate on what might happen if a river or stream were dammed or blocked by debris. Guide them to see that the course of the water would change, or, if the dam were strong enough, it would form a new lake.

Assessment: Children should trace the course of water on a map from a stream to an ocean or lake.

Modification: Children may wish to make their own water mazes.

Activity 4

Connection to Social Studies

The Air up There, page 24

Objective: Classify objects that fly or float in the air and objects that can't.

Introduce: Ask, *Have you ever flown a kite? Tell me about a day that is good for flying a kite. Can you fly a kite on a very calm day?*

Teach: Introduce the activity page and ask children to circle the objects that can fly or float in the air.

Close: Ask children if they can think of other things that fly, or float in the air.

Assessment: Children should correctly identify objects that fly or float in the air.

Modification: Children can draw an outdoor scene with some flying objects and some that can't fly.

Answers: (circle) kite, bird, balloon, plane, swallows, cloud.

Activity 5

Connection to Social Studies

Stop Pollution, page 25

Objective: Identify damaging forms of pollution.

Introduce: Show children two photographs, one of a clean beach or playground and one that is full of litter, broken equipment, etc. Ask, *Which one is a better place to play?*

Teach: Explain that different forms of pollution can spoil water, soil, and air. Introduce the activity and tell children to cross out the things that make Earth dirty.

Assessment: Children should recognize pollutants.

Modification: Children can color the rest of the picture with crayons.

Answers: (cross out) litter at the beach, car exhaust, smoke from the factory.

Activity 6

Connection to Art and Math

Play It Again, page 26

Objective: Learn how various items can be recycled or reused.

Introduce: Show photographs of old toys made from salvaged materials, for example, rag dolls made from old clothes, or products that are made from recycled objects.

Teach: Introduce the activity page. Explain that children will match objects to ways that the objects can be reused instead of being thrown away.

Close: Invite children to suggest ways to reuse other objects.

Assessment: Children should match the new objects to the objects from which they were made.

Modification: Give children some recyclable objects, such as boxes or empty bottles, with safety scissors, glue, and other craft materials, and allow them to make their own recycled toys.

Culminating Activity

Connection to Reading, Social Studies, and Language Arts

My Garden Book, page 27

Objectives: Use pictures to write about how a garden can be a resource.

Introduce: Explain that things in nature that we use to make food and other goods are called natural resources. Use the examples of milk from a cow and a tortilla from corn.

Teach: Show children how to fold the activity page to make a four-page book. Have them read the title aloud and study the pictures. Then ask them to use the title and pictures to write a story in the book about what would be in their garden if they had to get all the food they eat from it.

Close: As children share their stories, make a list of the resources found in a garden. Ask, *What would happen if we didn't have bees?* Help them understand that we would not have honey to eat.

Assessment: Children should be able to use the pictures in the book to help them write a story about natural resources in a garden.

Modification: You might have children work together to make a mural of a class garden.

Answer: Children's stories will vary.

Project Theme

Seasons and Snow, pages 28–34

Concepts
- Classify objects by weight.
- Use patterns to study how a liquid takes the shape of its container.
- Classify foods as solids or liquids.
- Write lyrics about gases for a song about matter.
- Predict whether an object will sink or float in water.
- Learn about measuring temperatures.
- Describe water in various states.

Overview In **Seasons and Snow**, children use the theme of a snowman and temperature to understand the concept of matter. Children use math skills to explore weight, patterns, and sequence of events; reading skills to explore cause and effect; art skills to apply what they know about matter; music skills to write a verse of a song about matter; and language arts and reading skills to write and apply vocabulary words. The unit culminates with a book to write about a snowman.

Getting Started Introduce **Seasons and Snow** as the unit theme. Lead children in a discussion of how water can be a solid, liquid, or gas. Discuss other objects that exist in different states.

Seasons and Snow Planning Guide

Grade 1 Unit E: Matter, Matter Everywhere	Activity	Resources	Related Subject	Macmillan/McGraw-Hill Programs	Materials
Chapter 9: Describe and Measure Matter **Lesson 1:** Properties of Matter	**Activity 1:** How heavy is it?, p. 28		Math and Art	*McGraw-Hill Math,* Chapter 11, pp. 413–414	objects of various weights
Chapter 9: Describe and Measure Matter **Lesson 3:** Liquids	**Activity 2:** Lemonade Time, p. 29		Math	*McGraw-Hill Math,* Chapter 11, pp. 411–412	bottle of liquid, bowls of different shapes and sizes, crayons
Chapter 9: Describe a and Measure Matter **Lesson 2:** Solids **Lesson 3:** Liquids	**Activity 3:** Lunch Time, p. 30		Social Studies	*My World: Adventures in Time and Place,* Grade 1, pp. 50–51	
Chapter 9: Describe a and Measure Matter **Lessons 2, 3 and 4**	**Activity 4:** The Matter Song, p. 31		Music		
Chapter 10: Changes in Matter **Lesson 6:** Solids and Liquids in Water	**Activity 5:** Sink or Float, p. 32		Language Arts	*McGraw-Hill Language Arts,* Grade 1, pp. 216–222	
Chapter 10: Changes in Matter **Lesson 7:** Heat Changes Matter	**Activity 6:** How Hot?, p. 33	*Temperature and You,* Betsy and Giulio Maestro	Math	*McGraw-Hill Math,* Chapter 11, pp. 427–428	
Chapters 9 and 10: Describe and Measure Matter; Changes in Matter **All Lessons**	**Culminating Activity:** My Snow Book, p. 34		Reading and Language Arts	*McGraw-Hill Reading,* Grade 1, pp. 108–119 *McGraw-Hill Language Arts,* Grade 1, pp. 216–222	

Scoring Rubric for Integration Activities	
Score	**Criteria**
4	Accomplished all of the activity's objectives.
3	Accomplished more than half of the activity's objectives.
2	Accomplished less than half of the activity's objectives.
1	Made little or no progress toward accomplishing the activity's objectives.

Activity 1

Connection to Math and Art

How heavy is it?, page 28

Objective: Infer the weight of objects by how much matter they contain.

Introduce: Ask children to name classroom objects. Ask, *Which can you hold in your hands? Which are too heavy to hold?*

Teach: Use classroom objects to demonstrate that all objects are made of matter, and the more matter an object contains, the heavier it is. Read the activity directions aloud.

Close: Ask children how they predicted which object was heaviest and which was lightest.

Assessment: Children should be able to make the association between weight and amount of matter.

Modification: Sight-impaired children can do this activity with classroom objects.

Answers: Circle around yo-yo; X on bicycle

Activity 2

Connection to Math

Lemonade Time, page 29

Objective: Use a pattern to solve a problem about the shape of a liquid.

Introduce: Demonstrate how a liquid takes the shape of the container it is in by pouring a bottle of liquid into bowls of different shapes and sizes.

Teach: Read the page directions aloud. Be sure children see that there are 12 squares in each pitcher, so they need to color 12 squares in each cup by starting at the bottom. Prompt them to think about how the liquid will look in the cup before they color.

Close: Draw a picture to illustrate the pattern made in each cup. Ask questions such as, *What do you notice about the pattern in each cup?*

Assessment: Children should use their understanding of liquids to explore patterns.

Modification: Some children might use blocks on a grid to explore the concept.

Answers: Children should color the bottom 12 boxes in each cup.

Activity 3

Connection to Social Studies

Lunch Time, page 30

Objective: Recognize the differences between solids and liquids.

Introduce: Review the properties of a liquid by having children name their favorite drinks. Then discuss how some liquids are poured on food, such as sauces, oil and vinegar.

Teach: Read the page directions before children begin the activity.

Close: As children share answers, help them contrast the differences between a solid and a liquid.

Assessment: Children should use their understanding of solids and liquids to identify foods that are solids and foods that are liquids.

Modification: Some children might need help identifying the foods pictured.

Answers: S: pizza, cupcake, salad; L: soda, punch, salad oil

Activity 4

Connection to Music

The Matter Song, page 31

Objective: Apply knowledge of matter to write a verse of a song.

Introduce: Review what children know about solids, liquids, and gases.

Teach: Lead children in singing "The Matter Song" to the tune of "Mary Had a Little Lamb." Then have children make a list of things they know about gases on the back of the activity page. Lead them in combining their ideas into a final verse such as:

> A gas is something you can't see.
> You can't see, you can't see.
> A gas is something you can't see.
> But sometimes you can feel.

Close: Display the verse the children wrote.

Assessment: Children should use their knowledge of matter to suggest ideas for a song.

Modification: Some children may need to dictate their thoughts to you rather than writing them down.

Answers: Children's suggestions will vary. Consider participation and understanding of gases in determining children's scores.

Activity 5

Connection to Language Arts

Sink or Float, page 32

Objective: Recognize things that float in water and things that sink.

Introduce: Ask children to share experiences of sailing toy boats.

Teach: Read the activity page aloud. Instruct children to circle things that would float and cross out things that would not.

Close: Invite volunteers to share their answers and to explain why they answered as they did.

Assessment: Children should recognize that certain objects will float while others would sink.

Modification: Ask children to draw some other things that would float.

Answers: (cross out) penny, heavy rock, book.

Activity 6

Connection to Math

How hot?, page 33

Objective: Read the temperature on the thermometer.

Introduce: Review how to read a thermometer.

Teach: Remind children that a thermometer is an instrument used to measure temperature. Have children complete the activity. Guide children to read the Fahrenheit scale.

Close: Display a classroom thermometer or draw a thermometer on the board. Ask a volunteers to read it for you.

Assessment: Children should be able to read the thermometer and fill in the correct answers: 80° F and 20° F.

Modification: Children who have difficulty with numbers may wish to work in pairs.

Culminating Activity

Connection to Reading and Language Arts

My Snow Book, page 34

Objective: Use the pictures to gather information to write about matter, specifically water, and its different states.

Introduce: Ask, *Have you ever built a snowman? What happened when the weather got warm?* (If children live in a warm climate, ask what happens to an ice cube that is left in the Sun.) Read a book about snowmen, then let the children use it as a reference for writing.

Teach: Help the children cut the page and fold it to make a 4-page book. Lead them in studying the pictures and describing the action. Then tell the children to use the pictures to write a story about a melting snowman.

Close: Have children make more illustrations for their snow books. Let children share their stories.

Enrichment: Challenge children to write or draw places in the school where they can find solids, liquids, and gases.

Assessment: Children should write stories based on information from the story and the unit, *Matter, Matter Everywhere.*

Modification: Some children may need to dictate their stories to you.

Answers: Children's stories will vary.

Project Theme

In the Playground, pages 35–41

Concepts
- Describe the forces of pushing and pulling.
- Apply position words to describe location and movement.
- Recognize magnetism as a quality of some kinds of metal.
- Identify missing parts of objects.
- Recognize that sound is the result of movement.
- Read and write words that describe motion.

Overview The theme **In the Playground** allows children to explore the concepts related to forces: motion, pushes, and pulls. Children use language arts skills to use a maze and to apply vocabulary words; reading skills to gather information from pictures; and art skills to draw missing parts of objects and to complete a picture. The unit culminates with a writing activity about forces at a playground.

Getting Started Introduce **In the Playground** as the unit theme. Begin a discussion of what children would expect to find at a playground, the equipment they have used at a playground, and how they have fun playing there.

In the Playground Planning Guide

Grade 1 Unit F: On the Move	Activity	Resources	Related Subject	Macmillan/McGraw-Hill Programs
Chapter 11: Force and Motion **Lesson 1:** Things Move	**Activity 1:** Push Land Pull, p. 35	*Pushing and Pulling,* Gary Gibson	Reading	*McGraw-Hill Reading,* Grade 1, pp. 152–187
Chapter 11: Force and Motion **Lesson 3:** The Ways Things Move	**Activity 2:** Where is it?, p. 36	*The Good Bad Cat,* Nancy Antle	Language Arts and Reading	*McGraw-Hill Language Arts,* Grade 1, pp. 91–107 *McGraw-Hill Reading,* Grade 1, pp. 227–289
Chapter 11: Force and Motion **Lesson 3:** The Ways Things Move	**Activity 3:** From Here to There, p. 37		Language Arts and Math	*McGraw-Hill Language Arts,* Grade 1, p. 91–107 *McGraw-Hill Mathematics,* Grade 1, p. 25
Chapter 12: Magnets and Sound **Lesson 4:** Things Magnets Move	**Activity 4:** Move It!, p. 38		Reading	*McGraw-Hill Reading,* Grade 1, pp. 227–289
Chapters 11 and 12: Force and Motion: Magnets and Sound **Lesson 3:** The Ways Things Move **Lesson 4:** Things Magnets Move	**Activity 5:** What moves it?, p. 39		Reading	*McGraw-Hill Reading,* Grade 1, pp. 227–289
Chapter 12: Magnets and Sounds **Lesson 7:** Moving Things Make Sound	**Activity 6:** Missing Pieces, p. 40		Art and Music	*Share the Music 2000* Grade 1, pp. 52–53
Chapters 11 and 12: Force and Motion; Magnets and Sound **All Lessons**	**Culminating Activity:** My Playground Book, p. 41	*Pushing and Pulling,* Gary Gibson	Reading and Language Arts	*McGraw-Hill Reading,* Grade 1, pp. 227–289 *McGraw-Hill Language Arts,* Grade 1, pp. 91–107

Scoring Rubric for Integration Activities

Score	Criteria
4	Accomplished all of the activity's objectives.
3	Accomplished more than half of the activity's objectives.
2	Accomplished less than half of the activity's objectives.
1	Made little or no progress toward accomplishing the activity's objectives.

Activity 1

Connection to Reading

Push and Pull, page 35

> **Objective:** Describe the forces of pushing and pulling.
>
> **Introduce:** Push a desk or chair gently across the room. Ask children, *Am I pushing or pulling?* Pull the object back in place, asking, *Now, am I pushing or pulling?*
>
> **Teach:** Read the activity page with children. Write the words *push* and *pull* on the board, and have children complete the activity.
>
> **Close:** Invite volunteers to share their answers.
>
> **Assessment:** Children should differentiate between pushing and pulling.
>
> **Modification:** Children who have difficulty writing may complete the activity orally.
>
> **Answers:** Child is pulling elephant and wagon; pushing snowball and car.

Activity 2

Connection to Language Arts and Reading

Where is it?, page 36

> **Objective:** Use position words to describe the location or movement of objects.
>
> **Introduce:** Review position words by reading a story such as *The Good Bad Cat*. Let children summarize the story as they identify the position words used.
>
> **Teach:** Explain how each word in the box will complete one of the sentences. Then invite children to do the activity.
>
> **Close:** Ask children to use each of the position words in a sentence of their own.
>
> **Assessment:** Children should use position words to describe the location or movement of objects.
>
> **Modification:** Play a game of Simon Says using position words.
>
> **Answers:** **1.** on **2.** across **3.** over **4.** in **5.** under

Activity 3

Connection to Language Arts and Math

From Here to There, page 37

Objective: Understand and use position words.

Introduce: Write the words *under, over, around,* and *through* on the board.

Teach: Introduce the activity page. Have children trace the correct path through the maze.

Close: Ask a volunteer to explain how he or she completed the maze: by following the monkey under the bar, over the rock, around the tree, and through the hoop.

Modification: Children may prefer to work in pairs.

Activity 4

Connection to Reading

Move it!, page 38

Objective: Recognize magnetism as a quality of some kinds of metal.

Introduce: Ask children why magnets will stick to the refrigerator but not to the wall or window.

Teach: Read the activity page with children. Invite them to complete the activity.

Close: Ask children to describe some other things that would or would not be affected by a magnet.

Modification: If children have difficulty differentiating between magnetic and non-magnetic objects, let them experiment on various materials with a small magnet.

Answers: (circle) nails, paper clips, pins.

Activity 5

Connection to Reading

What moves it?, page 39

Objective: Learn how to use terms to describe motion.

Introduce: Write the words *fly, turn, pull,* and *hit* on the board. Ask volunteers to use each of the words in a sentence.

Teach: Read the activity instructions with children. Have them work in pairs to finish the activity.

Close: Ask volunteers to read their answers.

Assessment: Children should correctly complete the sentences.

Answers: **1.** pull, **2.** hit, **3.** turn, **4.** fly

Activity 6

Connection to Art and Music

Missing Pieces, page 40

Objective: Recognize that sound is the result of movement (vibrations) caused by one object striking another object. Children should draw the missing parts that help produce sound.

Introduce: Display a simple instrument such as a triangle. Ask, children to describe how to use it to make noise. Ask, *Can the instrument make noise by itself?*

Teach: Read the activity page with children. Complete the first item with them or ask a volunteer to explain his or her answer.

Close: Have children draw another musical instrument.

Assessment: Children should supply the missing parts of the instruments in the illustrations.

Culminating Activity

Connection to Reading and Language Arts

My Playground Book, page 41

Objectives:	Use pictures to gather information to write about forces.
Introduce:	Read a book about forces, then use it as a reference for writing.
Teach:	Help children cut the activity page and fold it to make a 4-page book. Lead them in studying the pictures. Then have children use the pictures to help them write a story about what they would push or pull at a playground.
Close:	Let children share their stories.
Assessment:	Children should use the pictures and information in the book as a resource as they write their own story.
Modification:	Some children may need to dictate their story to you.
Answers:	Children's stories will vary.